AF822064

What's Inside

Quick Intro

Web developer, programmer and designer, Holly Valero has over 25 years of experience in all phases of website creation, maintenance and troubleshooting. She has built hundreds of websites and assisted countless individuals who were wrestling with corrupted passwords, lost domain names, mystery hosting, anti-social social media, hacked websites, funky code, and broken images.

The Pocket Webmaster is the collected strategy, smart ideas, organized approach and simple, easy-to-understand explanations that every website manager needs at their fingertips – whether you are an entrepreneur, or managing a web team, a nonprofit board member, or small business owner.

This operations-level guide covers what no other guide does: how to establish, manage and protect your digital empire's most valuable assets. Whether you are setting up your first online presence, or you've inherited somebody else's problem.

boilerplatebooks, llc
www.boilerplatebooks.com

ISBN: 978-1-7347934-3-7

Quick Facts: Domain Names

WHAT'S A DOMAIN? | HOW THEY WORK | HOW TO BUY | FOOLPROOF DOMAIN STRATEGY | WORKSHEET 1

TL:DR Human beings speak in words. Computers speak in numbers. Domain names – simple, combinations of letters or words – were created to assist with navigation when surfing the web. "**Popeyes.com**" is easier for us to remember than a 12-digit numeric sequence.

Underneath every domain is a hidden,12-digit numeric address based upon the geographic web server location and web hosting company. Every hosting company has at least one server that just keeps track of which name goes with which number. It's known as a **DNS server** (Domain Name Server) and acts as kind of a phone book, looking up domain names like popeyes.com and pointing visitors to the 12-digit numeric address associated with the website location. It also makes sure that what you see in your web browser is the domain name and not the number.

Domain names are like your real name. Everyone calls me, Holly, but it's just one piece of identification. We all have a physical address, mailing address, and a ton of other numeric tidbits like driver's license numbers, passport numbers, and phone numbers.

When you move, you have to update the address information stored by friends, family and the post office to point to your new physical location. When you change web hosting companies, you do the same thing for your domain name – for the same reason.

BTW: You can look up that number and geographic location of any domain name at websites like iplocation.net and this can come in handy if you're trying to figure out if your website is down for everyone – or just you.

There are lots of domain extensions. Some are reserved for government (.gov) or educational institutions (.edu) and every country has their own extensions like .ca for Canada.

When you buy a domain, you purchase it for a specific time period (1-10 years) and then renew it. Once you no longer need it, you stop paying for it and someone else can buy it.

TL;DR: Too Long, Didn't Read | **BTW**: By the Way | **DNS Server**: a computer that just manages domain names

How they work

If you enter HollyWorks.com into your web browser, the browser needs to match the name you entered with that hidden 12-digit IP number (192.254.225.164). This matching information is stored on those DNS servers and on local networks if it's a popular website that is frequently requested.

Your web browser is not going to verify Google's address every time you Google something. It's going to use the address currently on file. This speeds up surfing.

If you enter a domain name that your local servers don't know, then the request makes its way up the line to other Domain Name Servers until a match is made. Because computers are so fast, you don't notice a lag.

Once a match is found, you'll see **hollyworks.com** in your web browser and not the 12-digit number because your web host has both pieces of information and swaps the name in for the number.

How to buy a domain

FWIW: I recommend **buying domains from registrars**. Always. Always. Always.

Domain names can be purchased from either a registrar (GoDaddy.com, Domain.com, NetworkSolutions.com) or a **web hosting company**.

Your web hosting will probably change several times over the life of your website. If you buy your domain name from a hosting company and then find yourself switching to a new hosting company? You'll end up receiving two bills – **both** from web hosting companies.

You will **forget** that your **domain name** is still being managed by the old company, and will toss that bill thinking, *"oh, we're not with Rootin' Tootin' Hosting any more!"* Three months later, your domain will expire and you'll discover your website is gone – along with your domain. **You may never get that domain back.**

If this sounds confusing, just remember: **register domains with registrars. Buy hosting from hosting companies.**

BTW: Before you make a final domain decision, run your ideas ("SumoFoodTruck") through knowem.com to see if your name is available as a **domain**, and a **username** for dozens of **social media platforms**. If your marketing includes social media? A small adjustment may open up a world of marketing opportunities.

FWIW: For What It's Worth | **BTW**: By the Way | **Registrar**: a company that sells domain name registrations

Foolproof Domain Strategy

1. **Begin with a dedicated Gmail account:**
   Set up a separate Gmail.com account for your business. If you want to register **sumofoodtruck.com** create **sumofoodtruck@gmail.com** first. When that's done? Use that email address to sign up for domains and hosting so all notifications go to one predictable location, and don't get lost in your personal mail. If you become wildly rich and sell your food truck business? All the important website stuff is in one account that you can easily hand over to the new owner.

2. **Shop around for registrars:**
   I use **GoDaddy.com** but it doesn't hurt to compare (GoDaddy.com, Domain.com, NetworkSolutions.com.) Set up your account **first** using that Gmail address from the step above. Next, update account settings so you can login with a name like **sumofoodtruck** instead of a long customer number. Once this is all set and your credit card information is in there? Buy your domains last.

3. **Stick to the big three domains:**
   .com, **.net**, and **.org**. When was the last time you went to Google.xyz? You just want 1-3 domain names, registered for one year each, which should be around $35-$50. One-year registration makes for easier accounting, and forces you to check in once a year to make sure your data is up to date.

4. **Don't buy anything else:**
   Every online service – seriously. every. single. one. – will try to sell you everything from more domains, to hosting, email accounts and SSLs – blow by all of that. You just want the domain. I don't care if Eric in live chat tells you that their SSL certificates are way cheaper or their hosting is on sale. **Don't do it.** Remember, you can buy anything and everything else later.

5. **Avoid Private Registration:**
   Private registration sounds smart, but makes it impossible to look up your own domain details if you later forget them – specifically, the email address of the domain name owner. Again, you can always add it later. **Some registrars will apply privacy by default**. Check your listing at ICANN after you buy your domain, and if it's all private? Consider removing that privacy at your registrar for one variation of the domain. If you are using the .com? Remove privacy for the .net version.

OMG: Unless you're in Witness Protection I never recommend private registration. It is the **number one reason why people lose their domain names**. *It's also why we set up a dedicated email account in step one.*

OMG: Oh My God

Digital Assets Worksheet

Save yourself from a future panic attack and document your domain purchase here.

Make a note of the dedicated Gmail account and login, the domain name(s) purchased and login information and any extra recovery items like PINs, super-secret numbers or safe words.

WTH? Isn't it terribly dangerous to write down a password anywhere? Yes, but do it anyway. If this is the print version? Pop it in your safe, or store it with your organization's board materials that get reviewed yearly. If this is the fillable PDF version? Do not send it by email without **password protecting** the PDF first in Adobe Acrobat. And if you have a photographic memory? Skip this part...

Email Account:

Email Password:

Acct. Recovery items:
PINS, Customer number, etc.

Domain Name(s):

Registrar:

Username:

Password:

Phone or PIN:

Date of Purchase:

Date of Renewal:

WTH: What the Heck?

Quick Facts: Hosting

TL;DR Until you are ready to build a website, you don't need to set up a hosting account. Your domain names are safe **parked** at your registrar. Choosing a web hosting company is a bigger decision.

Things to ask first: do you need ecommerce or billing?

There's a difference between online sales and digital billing, or electronic payments. Your landscaper may have a basic website, but may bill you using pay-enabled invoices via QuickBooks. Your hairdresser may use Venmo or Zelle. Mobile apps like NerdWallet are becoming more popular. If you bill clients, check first with your accountant - or look for digital payment systems that work with your bookkeeping software. If you are not selling online? You don't need ecommerce.

If you plan to sell *tangible* goods with shipping, *print on demand*, digital product *downloads* or even *online teaching*, then the type of ecommerce needs to match the product or service.

Ecommerce platforms frequently come with **monthly fees**, **transaction fees**, or **both**. Many have tiered plans. Start by reviewing what's currently out there like BloggingWizard.com's 10 Best Ecommerce Platforms To Sell Digital Products In 2022. Click around and test drive demos and free trials.

If you do need some kind of ecommerce, what kind?

1. *I'm an author selling a book on Amazon!* You don't need an ecommerce website. You can embed an Amazon link at your regular website.
2. *I'm a craftsperson selling small felted animals for fun and profit!* Etsy.com may be all you need and you can link to it from your website or **redirect** your domain name to your Etsy store.
3. *I'm a T-shirt designer!* Check out Shopify.com. If you want to print designs on many items, try Teespring.com, RedBubble.com or Merch by Amazon.
4. *I'm a digital designer or artist or writer!* If all your products are downloadable, check out Gumroad.com, Sellfy.com, or SendOwl.com.
5. *I'm a Teacher/Trainer!* If you want to market online courses or webinars review Payhip.com, Teachable.com, or Podia.com.
6. *I own a food truck!* Research the industry as a whole at websites like FoodTruckr.com first. There are tons of options for payment systems made just for food trucks.

TL;DR: Too Long, Didn't Read | **Parked**: A domain you aren't using for an active website **Redirect**: A record that points your domain to another URL

Hosting and content management for non-commerce sites

Professional services, most industries, business-to business, nonprofits, advocacy groups, organizations, creative professionals, ministries and many other websites do not require ecommerce. Some need specific functionality like booking engines (hospitality) or reservations (restaurants) as an add on.

For these businesses, the web hosting decision is determined by your **CMS**, or **Content Management System** – the software used to create and manage your website content. Website developers work in a variety of programming languages and there are many content managers and some come with hosting.

Types of hosting, some bundled with content management

From the smallest/least expensive to the largest/most expensive:

Free website builders, all-in-one packages, industry giant freebies or add-ons:
Examples include **GoDaddy Website Builder**, **GoDaddy hosting**, or **Google Sites**. Unless your idea of a great time is being caller number 3,406 waiting for customer service, get hosting from hosting companies. If you need a professional website? This is not the option you want.

BTW: It's a well-kept secret in the technology field that only the **least competent** version of tech support is reachable by **phone**. "Wow, I got that fixed in one call," said no-one, ever. Look for smarter live-chat.

DIY Website builders:
Squarespace, Wix, and Weebly are the big three. On the plus side, they are very easy to get into and, if you have an eye for design and understand more than basic HTML, you can produce a small website by yourself. Downsides: you have to build every page one at a time. The fun factor fades quickly. If you enlist help from the platform it can be **surprisingly expensive**. Website templates tend to look better than the websites built using those templates. Because they cater to DIY webmasters, you can end up with a website with terrible SEO, or Search Engine Optimization. AKA, lousy Google search results.

TBH: You don't know what you don't know. Even with a good website builder, it's easy to spend a lot of time creating a website that – in the end – is inaccessible, low performing, and looks terrible on anything but your specific laptop.

DIY: Do It Yourself | BTW: By the Way | TBH: To Be Honest | CMS: server-based program used to create and display web pages

WordPress:
Is its own category because WordPress is enormously popular, with 43% of all websites running it. It's also the most hacked platform on earth, and requires **constant** vigilance, and weekly updates to prevent hacking. Secure WordPress hosting is more expensive, and may limit the plugins you can use, but you are less likely to experience hacking.

Everyone else? **You will be hacked**. You may not even know it. Fixing a hacked website can cost up to $10,000. Once cleaned, re-infection days or weeks later is common. Since WordPress websites tend to be on the same server at your web host, you can be hacked just by proximity to another WordPress website that is less secure.

WordPress is built upon the **core programming** (which does gets security updates) to which you add **plug-ins** (the cause of most hacking) and a **theme** for look and feel. All of these elements update regularly – but not **together**.

If your website depends upon a specific plug-in that's been hacked? Your entire website is at risk. The core security updates will **not protect** your theme or plugins, because there's no central integration. WordPress sites are notoriously slow because every plugin loads the resources it needs as if it was the only plugin. Five plug-ins that require jQuery (a JavaScript library) may load that entire program library five times.

SRSLY: If you value sleeping at night, do not build your website in WordPress unless you plan to include **secure WordPress hosting** (wordpress.com, WPEngine.com, Kinsta.com) or you have a full time WordPress developer who specializes in **security**.

Content Management Systems (open source) installed on a commercial server:
Joomla, Drupal, WordPress, Typo3, October CMS, Processwire, Fork, Grav, Butter CMS, Concrete CMS, and many others.

Some offer a hosted platform, like WordPress. Others will be installed by your web developer, which is why it's best to wait. Once you have a few website proposals, take a look at the CMS packages they recommend and review them for flexibility, extensibility, and ease of use.

Boutique Content Management Systems (bespoke) installed on a developer's server:
Truly custom web development is what you'll need if you have plans for an enormous content website like a national news publication, or you're creating your own web-based application, or your own version of an online community. If you have a huge web presence? You probably have a team to manage it.

SRSLY: Seriously

Hosting Alternatives to Avoid

What about leaving it up to social media?

Social media platforms like Facebook, LinkedIn, Pinterest, Twitter, Instagram, SnapChat, TikTok, and YouTube may be part of your business marketing strategy or customer/client outreach, but none of them can take the place of a professionally designed website. Social media should be fun. It should be fast and easy to use. But more and more? It isn't.

Facebook is the undisputed king of social media, but:

» Not everyone is on Facebook and Facebook requires a login.

» Google doesn't spider Facebook.

» You have little-to-no control over your content and website destiny.

» Your content can be removed and you can be blocked for a variety of reasons, and getting unblocked is not guaranteed.

» Organic exposure – the best part of Facebook ten years ago – is **no longer a thing**.

» Facebook's "Meta Business Suite" is not fun to use and takes a lot of time to learn.

» Ads are expensive and offer minimal rewards. Paid, boosted posts tends to kill organic exposure.

» Facebook has a bad reputation. And it's only getting **worse**.

What about running my own server?

If you love to tinker and have no actual plan of hosting a running website, knock yourself out. The Internet is an information superhighway. Your own web server is a private dirt road unmapped by GPS.

Some guy told me...

Apparently there's this guy out there who loves to offer opinions on everything. *Some guy told me that domain registrars are a scam and I should use a free public DNS service... He also said that big hosting companies are just trying to soak you for $3 a month when you could use this awesome server on the dark web for just .0000000001 bitcoin!*

If it was that great a deal? I'd be recommending it to my customers.

Foolproof Hosting Strategy

1. **Big, faceless hosting companies (BFHCs) with 24:7 tech support by live chat are better:**
   This seems counter-intuitive, but because BFHCs do not ever want to talk to you, they tend to provide more tools at the server and better documentation. Their software is updated constantly. Their downtime in microscopic. When there's a server problem, they swiftly move all websites on that server to a new server and you are unlikely to even notice. They don't take breaks. They don't take holidays. Yes, yes, it's terrible, but for your website? It's better.

2. **Think growth when you set up your first account:**
   You may be able to start with the tiniest hosting account, but it's better to aim for something in the middle. Tiny accounts may not include SQL databases, or limit you to one domain name. This doesn't mean you need a dedicated enterprise-level web server with a firewall. Shared hosting will probably work just fine. And do not buy a "website builder package" from any hosting company.

3. **Pick hosting that works with your chosen CMS**:
   For WordPress, that's secure WordPress hosting. For most Open Source platforms, you'll need a hosting package that supports MySQL, PHP, and offers server tools like *cPanel*, a server management UI that makes web management easier, from creating email accounts to adding subdomains. The better the tools, the faster and easier it is to develop your website.

4. **Avoid "one click installs" hosting:**
   Some web hosting companies provide lots of open source packages that can be installed with one click! While easy to install, they can be a real pain to work with or update. Don't let that drive the decision, and check with your web developer before you sign up.

5. **Ask your web developer what they need or recommend, but avoid hosting on their servers:**
   It can be super-awkward to break up with a hosting company tied to your developer. You may want to redesign your website. Maybe you don't love the CMS and want to try something new. Maybe it's just too expensive. Or, they take too long. Maybe you used to love them, but Eric left and now, it's just not the same... If you own the keys to the hosting account, you are in charge and you can change the locks.

LOL: In over 25 years of developing hundreds of websites, I have never offered hosting or domain registration. Maintaining web server software is just a little bit more fun than resetting passwords.

LOL: Laugh Out Loud

Web Hosting Worksheet

You're memory isn't as good as you think it is, so document your web hosting here.

Many hosting companies have similar sounding names: Big Host, Big Hosting, Bigger Host, etc. In addition to the name of the host, add their actual URL, and if they send you a super-secret URL to login as a customer? Make a note of that, too.

KIM: You frequently get a discount on your first hosting order, so it's good to sign up for a yearly renewal at the very least. The regular price will kick in after that. A great five-year hosting deal today may give you **sticker-shock** when it renews, so do the math first.

Hosting company:

URL:

Customer Portal:

Username:

Password:

Phone or PIN:

Customer number:

Date of Purchase:

Renewal Date:

Primary domain:

KIM: Keep In Mind

Quick Facts: Connecting Domains to Hosting

HOOKING UP YOUR DOMAIN NAME   TINY INTRODUCTION TO DNS MANAGEMENT

Why change DNS servers? What does that do?

A parked domain is one you own that isn't hosting a live website. Instead, it "points" to the registrar. If you pop it into a web browser, it will display a page from the place where you bought the domain, like GoDaddy.

When you create a website, you update the domain record to connect the domain name with the 12-digit address of the website at the web host. To do this we login to the domain registrar and either add the 12-digit shared IP address supplied by the web host (the A record), or, to make your life easier, just change those **two registrar nameservers** to the two nameservers that your hosting company sent to you when you signed up for hosting.

If you use the two nameservers provided by your hosting company, then when someone enters **sumo-foodtruck.com** into a web browser, your super-secret 12-digit address is provided by your **hosting** company instead of your **registrar**. Your hosting company also adds a ton of other details which means you don't have to worry about any of it.

I prefer this approach because it's easier, and troubleshooting always happens at the end where the website is displayed – the web host. If something isn't working, the hosting company can only fix what they can access. They won't login to GoDaddy. I like easy. So I just pop in the two nameservers and I'm done.

Hooking up that domain

YOLO: Take a deep breath. This is your first boss-level web encounter, but it's a small boss. Prepare yourself young Jedi, and may the Force be with you.

Your web hosting company will provide you with two DNS server names that look something like:

» **ns1.bighost.com**

» **ns2.bighost.com**.

That **"ns"** is short for **Nameserver**, and there's always two sequential numbers.

YOLO: You Only Live Once

HOOKING UP YOUR DOMAIN NAME TINY INTRODUCTION TO DNS MANAGEMENT

When you login to your GoDaddy Account and painfully make your way to "**Manage DNS**" you will see a section called **Nameservers** and there will be two entries already there. Those are the DNS servers for your registrar. Somewhere you should see a button allowing you to *change the default nameservers*.

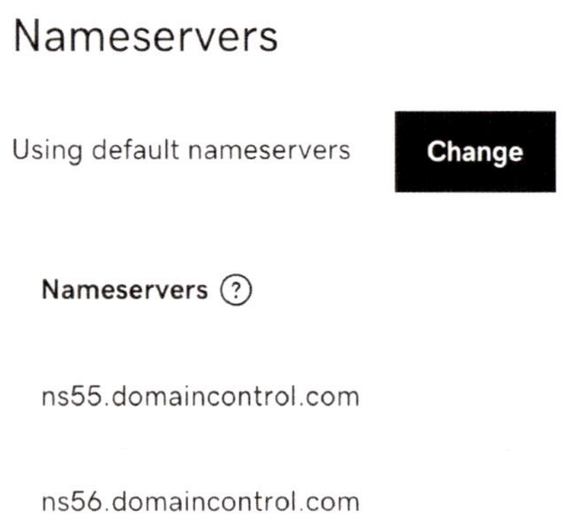

When you click "change," you will see the first of several threatening screens advising you that changing nameservers is the beginning of the end. Technically, we don't have a website set up, so nothing bad can happen. Select whatever is closest to: "**Change**," and **"Enter my own name servers (advanced)."**

Again, you are likely to see several warning messages. Ignore those and when you get to a form that allows you to enter your own nameservers, **pop in the two you got from your web hosting company and hit save**.

DFTBA: Take a moment here to reflect upon your courage, bravery and smarts. You've just dipped your toe into DNS management! On to **propagation**.

What's propagation?

When you make a change to your domain nameservers, it takes time for that information to be widely distributed. If you buy domains at **registrars**, the change is usually half an hour. If you bought it from anywhere else, it will take longer – up to a couple of days.

Fun fact: it will take longer for you to see your own website live because all your local DNS servers around you have cached (or saved) the previous DNS address on file – the registrar's address. You have to wait for that data to be replaced with the new address.

FOMO: Always do a soft launch of your website quietly a few days before you announce it to everyone on earth. Flip the switch on Friday and it has the whole weekend to propagate. By Monday, you can be 99.9% sure everyone will see your new website.

Ready to host a website? Visit the **registrar** first, and edit your domain (**sumofoodtruck.com**)

The Hosting Path

This path is the easiest and most logical if you are hooking up a new domain, or you have basic domain needs: website, email.

Your hosting company sent an email with their 2 nameservers listed, but you can always look it up by logging into your hosting account dashboard if you forget.

Example:

ns1.hosting.com
ns2.hosting.com

Return to the registrar, login and **change their 2 nameserver entries to your hosting company nameservers.**

You're done.

Within 30 minutes the domain is live.

All domain edits are done at the hosting company from now on.

- MX records for email
- TXT records when necessary

The Registrar Path

This path is the easiest and most logical if you already have a pile of domain name additions for mail servers or TXT records, or a larger operation in general.

You will need the numeric address associated with your account, probably listed as a **Shared IP Address** from your hosting company.

Example:

154.233.122.421

Return to the registrar, login and look for advanced DNS editing or zone editing for the domain. Once there, update the A record with the numeric address from your web host.

You're done.

Within 30 minutes the domain is live.

All domain edits are done at the registrar from now on.

- MX records for email
- TXT records when necessary

FOMO: Fear of Missing Out

DNS Management Made Easy

While you were at your registrar updating those two Nameservers, you might have noticed a page with various bits of data about your domain name. In the same way that you have a name, but also a phone number, street address, and social security number, your domain name is much more than just a name.

All the letters and acronyms look scary at a glance, but in fact they are all pretty simple. You already know that **NS** means "nameserver" and it either points to your registrar, or your hosting company's DNS nameservers. The **A record** under "Type" points to that actual **12 digit hidden number** mentioned previously. The **CNAME** record points to another name. For the moment, that's all you need to know.

DNS Records

DNS Records define how your domain behaves, like showing your website content and delivering your email.

Delete   Copy   Filter   Add   ...

	Type	Name	Data	TTL		
☐	A	@	Parked	600 seconds	Delete	Edit
	NS	@	ns55.domaincontrol.com.	1 Hour	Can't delete	Can't edit
	NS	@	ns56.domaincontrol.com.	1 Hour	Can't delete	Can't edit
☐	CNAME	www	specialistpolo.com.	1 Hour	Delete	Edit
☐	CNAME	_domainconnect	_domainconnect.gd.domaincontrol.com.	1 Hour	Delete	Edit
	SOA	@	Primary nameserver: ns55.domaincontrol.com.	1 Hour	Delete	Edit

Quick Facts: Email

BEST PRACTICES FOOLPROOF EMAIL STRATEGY WORKSHEET3 MX RECORDS

There are four basic ways to approach managing email:

1. **Google Workspace:**
   Handy for individuals and smaller offices. Gmail is pretty easy to set up initially, easy to check on any device, is never down for any length of time, filters spam effectively, and accommodates people who keep an empty inbox as well as email hoarders.

2. **Outlook Mail Server:**
   Good for larger offices with a tech support team to administer it.

3. **Basic email accounts at the host:**
   Requires more knowledge on an ongoing basis to set up accounts with your web host, reset passwords, and troubleshoot missing mail. Most web hosting companies provide a few webmail programs to check your email at a server address like **mail.sumofoodtruck.com** and this can be helpful when troubleshooting situations like, "I never got your email," since you can see if the message at least made it to the email server.

BTAIM: It takes little time to max out your server space if you leave email on your web hosting server. If you decide to change hosting companies? Trying to move all that stored email off the server is a real drag. So use webmail to troubleshoot, but check accounts using a mail program that won't leave mail on the server.

4. **Using another email address:**
   There was a time when a non-domain email like **sumofoodtruck@gmail.com** was viewed as unprofessional and a little sketchy. Times have changed. If you are a sole proprietor, or a small business it's okay to use that Gmail account.

ROFL: That doesn't mean you should use an AOL account, or an unassociated or confusing address like **batmanlives@ravinglunatic.com**. Think professional.

BTAIM: Be That As It May | **ROFL:** Roll On the Floor Laughing

Foolproof Email Strategy

Email is a no-brainer until it stops working; at which point it becomes a national crisis. The best strategy for email is to match your approach to the people who will be using that email.

1. **Think first about the kind of accounts you want to create:**
   Eric@hotpizza.com only works while Eric does. But, if you're Eric? And it's your pizza place? It's perfect. F.Jameson@stuffylawyers.com is better than Frankie@stuffylawyers.com because it sounds more professional and accommodates a larger employee pool before you need to add a middle initial. Historically popular approaches like "customerservice@facelesscorp.com" or "info@anything.com" work fine for very large organizations, but for smaller firms they tend to give the impression that your mail will go unanswered. Funny addresses sometimes work (thejunkguy@oldtruck.com) and sometimes don't (stiff@morrisfuneralhomes.com.)

2. **Fewer accounts is better:**
   Checking multiple accounts is more work than it should be. If you are creating extra accounts or email aliases to try and pre-sort your email, maybe combine them into one customerservice@shop.com instead of: *neworders@shop.com, shipping@shop.com, subscriptions@shop.com*. Start with fewer – and add on as needed.

3. **Email aliases can be flaky, so avoid them:**
   An **email alias** looks like an email address (info@whatever.com) **but doesn't have an actual account that can be checked**. Instead, anything mailed to that account is **forwarded** to another actual account. This was a very common practice for years and made it easy to look like a legit business email address, while conveniently sending mail to your personal address. This practice is ending, and the result is that forwarded email aliases are much more likely to get lost or never delivered. And **because there's no actual account?** You can't troubleshoot your aliases by checking webmail! Fortunately most email programs will let you check multiple accounts. And if you use Google Workspace as your email program, you can set up aliases within Gmail that will get delivered.

4. **January is unsubscribeuary:**
   In the average year I send over 2,500 emails that take up over 1gb of space. Google Workspace offers a generous 30gb allowance, but it's a great habit to prune and purge your email. I save sent mail for five years. Every January, I spend the month reviewing every ding-dong mailing list I signed up for and ruthlessly unsubscribing. I view my sent mail by reverse date (oldest first) and delete anything over the five year mark. Businesses routinely send PDFs and all sorts of big documents by email. It adds up. Don't pay for extra storage if you don't have to.

Email Alias: A fake email address that forwards mail to a real email address.

Email Game Plan

If you end up selling your Sumo Food Truck business, or move on to a new nonprofit, imagine how easy the transition will be if you keep track of all the important stuff in one location.

HT: You get bonus points for indicating here the person who is checking that business Gmail account we created back in chapter one.

Who is checking the Gmail business account?

Email is managed by: Google Workspace Outlook Mail Server Web Host Gmail Account

Who is serving as email administrator for Google Workspace or Outlook?

Who is in charge of setting up new accounts or resetting passwords?

What naming convention are we using? (Provide an example)

Starter accounts:

HT: Hat Tip

Email MX Records Made Easy

TL:DR: If you are using an account like sumofoodtruck@gmail.com, or setting up a couple of records at your web hosting location (Eric@sumofoodtruck.com) you don't need to worry about MX records. If you have a tech support team handling your mail via Outlook, then this is not your problem, either, so pass the buck.

MX is short for **Mail Exchange**. In the same way that your physical mail makes it to your apartment or a post office box, your MX records spell out where to send, receive, or - more simply - **exchange** mail. If you use Google Workspace to manage your mail, you'll first need to edit your DNS record for your domain (again,) and add all the MX records for Gmail. There are five addresses. The first item is the name of Google's mail server (aspmx.l.google.com) while the second item indicates its priority:

aspmx.l.google.com	1
alt1.aspmx.l.google.com	5
alt2.aspmx.l.google.com	5
aspmx2.googlemail.com	10
aspmx3.googlemail.com	10

OMG, do I have to do this for every email account?

No. You set this up once for the domain, no matter how many email accounts you have. Setting up Gmail at Google Workspace also happens once for the domain, and you create and manage all email accounts at Google Workspace.

Why are there so many MX entries?

For the same reason big cities have more traffic lights. Lots of traffic. Over 300 billion emails are sent every year. That's a lot of mail. Starting at the top, when mail is sent to you, the message hits the first server on the list with the number one priority. If it's too busy, it makes its way to the next, until it finds a server that's free.

TL;DR: Too Long, Didn't Read

How do I add MX records?

When we connected our domain name, we used the *two big DNS entries from the hosting company*. That means, all the rest of our DNS management happens at the **hosting** end. So that's where we are heading next.

DYK: What we're about to do may look scary? But it's not. If we are setting up a new domain, you aren't even getting email yet, so you can't break anything.

When you login to the customer portal at your web hosting location there will be a lot of information, but it's usually broken up into predictable sections like "mail" and "domains." Look for the section for domains since we are setting up email for the whole domain, not creating or checking an email account. Look for **advanced DNS editing** or **zone editing**.

My web host is Hostgator.com, which uses cPanel (kind of like a little operating system for your server to make things easier) so once I locate the **DNS Zone Editor** I can choose the **MX Record button** for any domain. cPanel makes this easier with a little pop up window. It reduces copy/paste errors, and minimizes the chances that I will need to call tech support. See "Scary Screenshots."

Since Hostgator is a **BFHC**, and never wants to talk to me? It's a win-win.

But even if your web host doesn't have a specific tool, you can add MX records directly in the zone editor. Just copy and paste carefully from the list of Google MX servers. You'll choose **MX** for the type, and note the matching priority number for each server.

When you are done adding MX records, double check for the stray **default** MX record that you didn't add. There's no doubt one in there as a placeholder. Delete it to keep email from having a freak-out.

NBD: Still terrified? Take a **screenshot of your DNS record** (command+shift+3 [Mac] or control+shift+3 [PC]) before you start. This way if you make a "terrible mistake," you can check the screenshot for what was there originally, and, if necessary, add it back.

TTL stands for **Time To Live**, it's kind of like a stopwatch that notes in seconds the amount of time your data is saved before a fresh query is made. The default TTL is expressed in seconds: 14,400 seconds is 4 hours and that's the default for most MX records. This doesn't need to be changed.

DYK: Did You Know | **NBD**: No Big Deal

Scary Screenshots

Maybe not-so scary. This is just your **DNS Zone Editor**. Each "zone" governs a specific function like mail or your website.

Find the domain first, and if there's a handy **"+ MX Record"** tool try that. If not, just edit the domain record and add MX entries one at a time. Your **valid zone name** is your domain name followed by a period. It just has to be there and there's a super-nerdy reason why, but it's not interesting enough to explain, and you will never need this information for anything else.

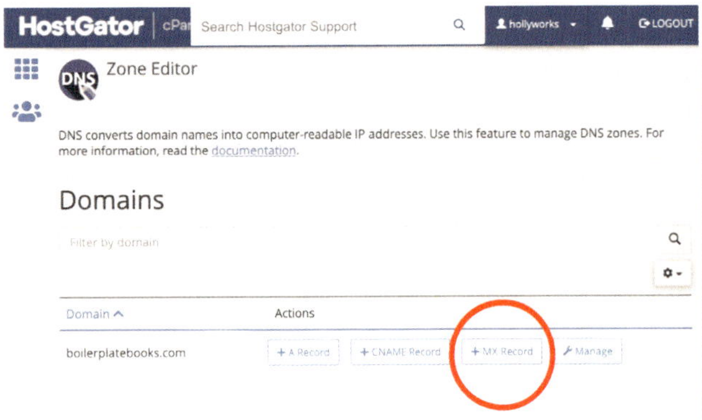

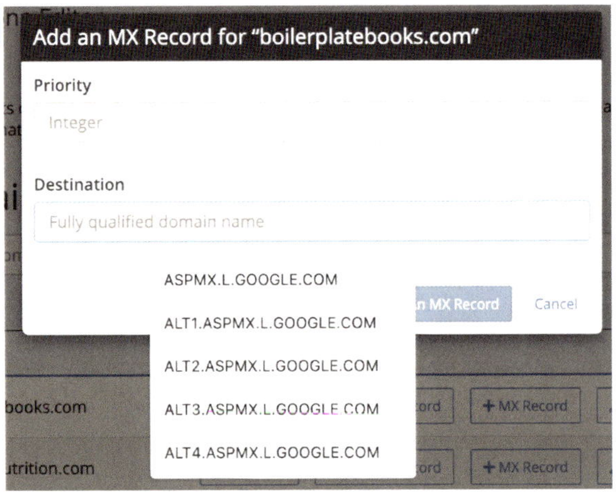

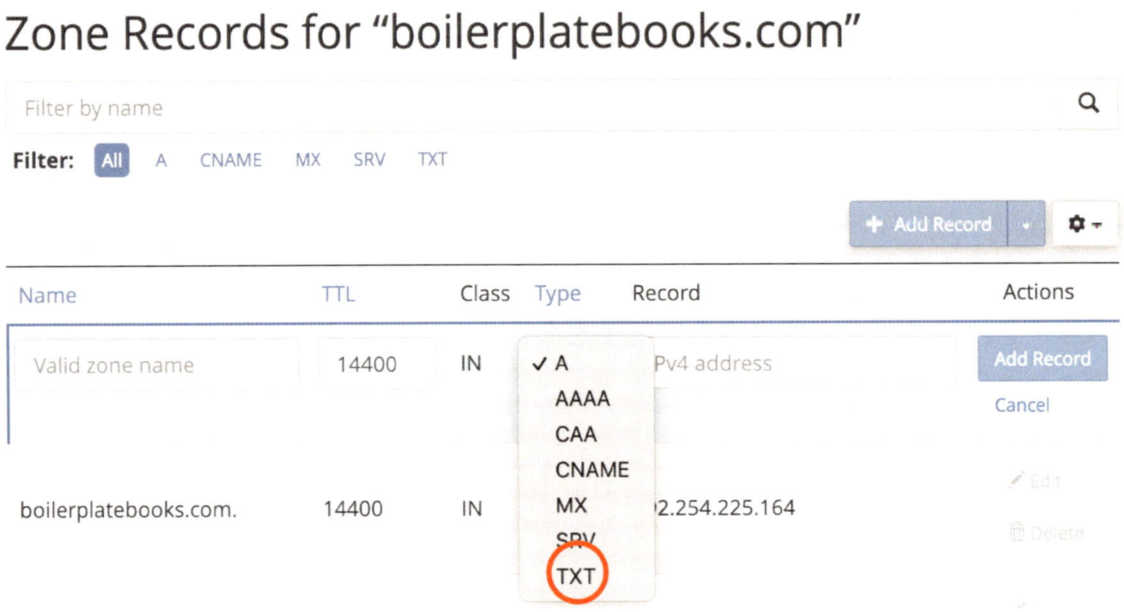

Quick Facts: Social Media and Third Parties

NAMING STRATEGY SECURING ACCOUNTS VERIFICATION WORKSHEET 4

You've purchased one or more domains, reviewed CMS options and hosting, and possibly signed up for a hosting account. We've made a game plan for handling email.

Securing your online identity comes next. Not every business wants or needs social media, but it's still a good idea to come up with one or two predictable user names that you can use for everything from third party services like reservation engines, to YouTube.

Since most accounts have a character limit, shorter is better. Things to avoid? Unnecessary clutter (llc, esq, assoc, inc, corp, online), acronyms for super-long names (DOGRPXLLP), selling phrases that don't say anything about you (@buyamug) and numbers–unless you're a professional athlete.

Come up with one or two logical options, then do a quick search of Knowem.com and Namechk.com to see if your name is available across several different platforms. A single search reveals that: sumofoodtruck is available just about everywhere.

These services offer sign ups as well, but unless you plan on hundreds of accounts, it's better to **sign up for accounts yourself**. Set them up now, before you plan to use them, while they are still available. The obvious accounts are LinkedIn, and any social media you plan to use like Facebook, Instagram, Twitter, YouTube, etc.

OTP: This name is **public-facing** and shows up in URLs, so avoid anything clueless (@stupidcustomers) or branded poorly (@ohiosocial_client6) or dumb (@erics_fbacct.)

OTP: One True Pairing

Securing Your Identity

Even if you have successfully set up your business email, it may still be better to use sumofoodtruck@gmail.com to set up third party accounts where you can. That address will never expire, and will still work even if your domain name expires. (Which it won't–right!?)

Some social media platforms like Facebook require personal accounts to get in, so for any like that, use your personal email. You can always add or remove yourself from any business page. You may be surprised at how many services you end up signing up for over time: cloud storage, CRMs, donation management, form engines, ecommerce, email services like Mailerlite or Constant Contact.

The Internet service industry is exploding. Most important is to **document** everything as you add them:

» Usernames

» Email address associated with the account

» Password

» Agency or individual managing this

» Name of the company and it's URL so in the event that company is purchased by a larger firm, you can trace the path to the current owner

» Freshness date (for example: *password is RHINO810xpq as of May, 2021*)

Avoid at all cost:

We had this great intern last summer named Lauren, and she set up a bunch of social media accounts for us. I think she's in Tennessee, now, but I'm not sure. I don't know what happened to our Facebook page. I guess it's still "there."

Make sure that you are in control of your social media. You can set up a personal Facebook account, and then a Facebook business page and make Lauren a *contributor*. Social media is the front lines for reputation management. Guard it with your life.

WCS: Think I'm over-reacting? Okay, Lauren leaves – and leaves Facebook. Your Facebook business page is officially cast adrift. Months later, Russian trolls hack her neglected account, and use your page for nefarious purposes. You receive tons of complaints, but you can't get into the page to fix it.

WCS: Worst Case Scenario

Validating Ownership: .TXT Records

Back in the good old days you could sign up for anything online and everyone just took you at your word. Not anymore.

Remember how we bought a domain, and then plugged in those two DNS nameservers? And then I made you go back and add a bunch of MX records? I did that because the modern day web is much more inventory management and far less Magical Mystery Tour. This brings us to another DNS entry known as the TXT record.

TXT records don't actually do anything. They are added to the domain record so companies like Google can verify domain ownership. How? You sign up for a Google service and Google gives you a TXT entry that's a long alphanumeric scramble. You edit your DNS record (at the web host in this case) and add the entry, marked as a TXT. Google looks up your domain and finds the corresponding entry which verifies your account.

They get added usually one at a time, when you add some new web service. Because they don't actually do anything, you cannot break a website by adding TXT records.

```
yourdomain.com.    14400    IN    TXT    FRVCJNoqDZgAlxSjOsr7fP2KazLdWy
                                         vGnepXMb58496humi0kTwEUct1QYl3Br
```

Looking above, you'll see your **domain name** followed by the period (don't ask) and a **14400** which is 4 hours displayed in seconds. The cryptic "**IN**" stands for Internet – which you may have guessed. **TXT** is because it's just **text**. And this long scrambled number acts like a super-secret password between you and whatever service you are using. Over time, you'll end up adding upwards of half a dozen TXT records to your domain.

JIC: Notice I haven't been documenting these items. That's because you can look up your domain records at any time. If you look for these details at the **registrar** and don't find it, it's at the **web host**. It can only be in one of two places. What we are documenting is account access, and who has it, because that's not written down anywhere.

JIC: Just in Case

Creating a Services Registry

This is a job that really calls for a Google spreadsheet. Login to that dedicated Google account (sumofoodtruck@gmail.com) and make a copy of my Pocket Webmaster: Digital Asset Manager (it's public) and add it to your Google spreadsheets. Or, follow this sample if you'd like to set up your own.

Sample third party documentation

Company Name

Category

Description

Date Purchased

Date Expires

Auto Renew?

Email Address

Role

Username

Password

PIN/ Account #

URL

Troubleshooting: Inheriting a Website

A TYPICAL SCENARIO HOW TO FIND OUT WHAT YOU DON'T KNOW HELP IS MY WEBSITE DOWN?

Small businesses and nonprofits have smaller staffs, rely heavily upon volunteers and interns, and turnover hurts more. Nonprofit boards represent many levels of tech competency. Members also change every couple of years. Because everyone is doing the work of two people, no one has time to document anything. There's a false sense of shared knowledge – just ask Sarah!

But then, Sarah leaves...

The "new" person finds a completely undocumented digital collection of social media accounts, websites, services, and virtually no idea of the scope, or how to access any of them. Accounts logins were created over time with the email addresses of now-former bookkeepers, retired office managers, and part-time assistants. No one knows who set up what, or which email address was even used. Credit card numbers on file begin to expire. Emails reminding you to renew your domain are sent to sarah@longgone.com and when there's no reply? The domain expires.

Because of that nifty way local servers cache your data immediately around you? **You will be the last to know when it all falls apart**.

FTW: Remember this? *Even if you have successfully set up your business email, it may still be better to use sumofoodtruck@gmail.com to set up third party accounts where you can. That address will never expire.*

Every web service requires an email account. With access, any password can be easily reset within a couple of minutes. Without it (and now, a phone number) the process requires everything from faxing notarized documents to engaging in some days-long ritual. You can lose your domain name, your website, access to your email, and access to all your third party and social media accounts. It can take months to fix.

This is why it's critical to know the email account, be able to access it to reset the password, and know who was connected to the account. If it wasn't documented for you? You will have to piece things back together.

FTW: For the Win

Domain Sleuthing

You know you have a domain, but who bought it? When? From where? Don't panic.

The WHOIS search reveals all – maybe

Start by doing a "whois" search for your domain details at ICANN. You should see the **nameservers**. There will be two and they will either reflect the registrar or the web hosting company.

Hostgator's nameservers include **hostgator** in the name. But what if the nameservers are not that obvious?

Try a quick Google of what you find. For example, "whois domaincontrol.com" tells you it's GoDaddy's nameserver. I don't need the full domain record for them, and that's why I just Googled it.

While we're at the top of the record, notice the **dates** section? Look for **registry expiration** at the top of the list. This is when your domain name **expires**.

Domain Information

Name: HOLLYWORKS.NET

Registry Domain ID: 1538848689_DOMAIN_NET-VRSN

Domain Status:
clientDeleteProhibited
clientRenewProhibited
clientTransferProhibited
clientUpdateProhibited

Nameservers:
NS6237.HOSTGATOR.COM
NS6238.HOSTGATOR.COM

Dates

Registry Expiration: 2023-01-23 13:12:29 UTC

Updated: 2022-01-24 15:19:26 UTC

Created: 2009-01-23 13:12:29 UTC

I have a Google calendar just for important stuff like this and when I come across an important date, I immediately head to my calendar, and make a note for one-week prior to the expiration date. I include the link to the registrar, the name expiring, and anything I need to quickly renew that domain.

ELI5: Remember that dedicated Gmail account we created? Ten points to Gryffindor if you pause to add a Google calendar to that account **right now** and add your domain expiration date. I'll wait...

That is a **ton of life-saving information**. If the domain name doesn't have privacy protection invoked, you'll also see who bought it, what email address was used, and everything else you need to get back on track. If, however, the domain was purchased with privacy protection take a couple of minutes to swear or cry, and then make your way to the registrar listed and follow their steps to reclaim your domain.

ELI5: Explain It to Me Like I'm 5

Where to Find Help

When bad things happen people tend to panic (not good) and call someone (even worse) and after being on hold for an hour they are on the verge of a nervous breakdown. First, never call. Live chat, text or email is better. Save "talking to a real person" for your mom, your significant other, or your **BFF**.

Next, find out what you can (see **"Is My Website Down?"** next page) and then contact someone:

Webmaster: This person manages your website content. They may have built your website, or they may work for you updating content, assessing traffic and marketing, and possibly managing social media. If there's a problem with content at your website, check with your webmaster first. If there are problems with website display? Start with the webmaster to find out what's up.

Web Designer: This person may have built your website, or assisted on the team of people building your website. They may specialize in UI (user interface) or UX (user experience) or design and responsive design. Responsive design is the current standard for designing websites that are attractive and readable on a wide variety of devices and screen sizes from mobile to desktop. They "respond" to changes in width and display, by reflowing the content in the page instead of simply making web pages appear smaller. Contact your web designer if you're interested in a site redesign, or the overall design is producing errors.

Web Programmer: This person may have provided customized programming for your website. They may create MySQL databases, program in PHP or any number of languages. Web programmers often work as part of a team with web designers. Contact your programmer if you need to change or extend the current programming at the website, or if a specific program function isn't working correctly.

Web Developer: A web developer has the widest variety of skills. They may design, program, and create websites. They usually have more server expertise, and more advanced DNS management skills. Sometimes referred to as a full stack developer, a web developer can manage and oversee every aspect of website design and development. If you're working with a web developer, you can contact them about anything.

Hosting Company: Where your website lives. If you have problems with lag, websites timing out, and it's not just you? Contact the hosting company to see if there's a problem with the server.

Registrar: Where your domains live. If you try to visit your website and see an ad for domains? Or nothing happens? Or you see a different web page? Login to your registrar and make sure your domain bill hasn't lapsed.

BFF: Best Friend Forever

Is my website down? Is it me? The ultimate chart

I went to my website and nothing happened!! **That's not possible**. Something did happen. Let's find out what... Don't follow a bookmark or shortcut. Manually enter your domain name into a web browser.

What do you see?

Blank screen, or a spinning load icon

Make sure **you** are online. In another browser window, head to Google and search for "penguins" or visit: amionline.net

No penguins? **You're offline.**

Okay, visit any of these to see if the website is down - or just down for you:
doj.me
isitdownrightnow.com
downforeveryoneorjustme.com

It's just me. **Wait a bit.**

OMG, it's down!
Refresh your browser. Still blank?

no

yes

Your website may be fine, but there could be a piece of Internet equipment that's down somewhere between you and your website. If you're tired of waiting:

Visit network-tools.com and choose "**Traceroute**" from the dropdown menu and enter your domain name.

Visit the **Internet Traffic Report** if you know your website's geolocation and want to investigate further.

internettrafficreport.com

I see an advertisement

Is it for a hosting company?

no

yes

Your domain name is probably fine, but **your hosting has expired**.

Login to your hosting company and update your credit card data to **restart hosting**.

Is it for a registrar?

yes

Your **domain name has probably expired**.

Login to your registrar, and **renew your domain name** by updating your credit card.

Login to your default gmail account, go to **spreadsheets** and update your **Pocket Webmaster Digital Asset Manager** with the new date, and then update the **calendar** with a renewal reminder for next year.

I see an error number

500?
It may just be a server reset. Refresh a couple of times.

400?
You are looking for a page not there. Back off the URL to just the domain: .com/ not .com/index.asp

300?
This means you're being redirected to another page. Again, back off the URL to just the domain name without a page reference.

Hot mess?
Broken images, broken elements, lag, partial loading. Wait a couple of minutes and try again. If it's still a hot mess, your website is **broken**.

Contact your webmaster.

I see a foreign website or a different website - not mine

If you see a pile of advertising and have **no idea what the website is for?** Your domain name may have expired and it may have been sold to a bulk registrar.

Bulk registrars buy domain names from big regsitrars and hosting companies. It's less common today, and less likely to happen if your domain name expired recently. If it's been more than six months? Getting that domain back may not be possible.

If you see signs of malware, pornography, or anything else? Your website may have been hacked.

Contact your web hosting company and your webmaster.

Traceroute
Traceroute literally tracks all the computers and routers between you and your domain name. It's a long list of those 12-digit machine addresses, one after another. If a piece of equipment is down between you and your website, you will see entries like: *** *** ***

Setting Yourself Up For Success

DIGITAL ASSETS MANAGER FRUSTRATION TAKES A HOLIDAY SCAM-PROOF YOURSELF WEB TOOLKIT

90% of genius is just organization

I like to assume that my future self is prone to panic attacks, memory loss, and crying jags. So today, while I have my act together, I do everything I can to set my future self up for success.

Remember that default Gmail account we created for sumofoodtruck@gmail.com? That Google account became the first cornerstone of our success. To that we added a Google spreadsheet and a Google calendar – all in one Google location that, combined, became our Digital Assets Manager.

Digital Assets Manager as lifesaver

If you subscribe to all web services, domain purchases and hosting using the dedicated Gmail account, it's easy to scan your inbox for renewals, password resets, or anything else. It's the only mail that will be in there.

As you sign up for YouTube, Mailerlite, etc., if you get into the habit of adding each to the spreadsheet over time you'll end up with a brilliant reference tool that will make your life 90% easier.

At some point I began adding freshness dates to my spreadsheets because password changes happen, and if I get a notice from some website that my data been compromised, I may quickly change a password on the fly – and promptly forget about it. Staring angrily at my spreadsheet later when I can't login, didn't give me enough context. By adding the date, you have a better sense of whether it has just been changed.

Genius-level back up plan:

» If you're a small business, download the spreadsheet once a year and keep it with your taxes for that year.

» If you're a nonprofit, make the spreadsheet a part of your board agenda to review and document once a year at least. It's a great way to gently disseminate knowledge over time, and an opportunity to review any services you want to keep or toss.

Frustration takes a holiday

In your Digital Assets Manager Google calendar, I recommend creating a dedicated "**web stuff**" calendar and documenting everything like yearly domain or hosting renewals – any service. Add a quick link to pay, or check your credit card dates.

KK: An entry like **"renew domain!!!"** no matter how many exclamation points, is not helpful. Set yourself up for success with something more like: "HollyWorks.net autorenews at Godaddy.com for one year March 17, 2023. Privacy is off"

When you manage a website you regularly encounter challenging situations. Your favorite web tools are updated and your muscle memory no longer works. Because the Internet evolves in waves, you can experience months of calm, or months of complete frustration.

I needed a way to document anything that took more than ten minutes to figure out, or anything I knew I'd have to do again – someday. I added a monthly holiday to my Google calendar near the end of each month, on auto-repeat, called "**Frustration Day**."

Anytime I find myself saying, "well, THAT was a nightmare," I add the detailed short version to that calendar date description with any links I might need.

WDYMBT? A recent printer purchase included the option to email documents to the printer to print automatically when you're out of the office. Cool! I just needed to create a *super secret printer email address* within the printer software. No matter how logical or obvious my choice, the chance of me remembering it on the road is zero. In the Google calendar, I can look it up on **Frustration Day**. And, yes, I can add it to my spreadsheet, but it's not exactly web stuff, so the chances aren't great that I will think to look there while I am crying in my car. Either way, set yourself up for success.

Scam-proofing your web empire

Many clients forward emails to me that are scams with the question, "do we have this? Should I pay this?" Scams are everywhere and they can be very convincing. Here's a simple way to avoid most of them:

1. **Anything that arrives as physical, paper mail can be tossed out.** If you signed up for a domain name from GoDaddy.com, there's no way on earth you will receive a paper bill.

2. **Any phone call can be ended and the number blocked**. Google will not call you. Your hosting company will not call you.

3. **Any advance warning of disaster is a scam** and can be ignored. When have you ever gotten advance notice of something terrible? Never. Your web hosting or domain expires and you find your website gone. Not almost gone. Gone! Any call or email warning you that your **[whatever]** is about to **[whatever]** is crap.

4. **Never click anything in an email that stresses urgency**. Email scams love to suck you in with subject lines declaring that your Beyonce tickets are in the mail! It may advise you to click or call. Pause for a moment. Go to your dedicated gmail account and look for this service in your spreadsheet or emails. If the service is one you use, open a browser window and login to your account and look around. Toss the email. It may help to Google: "[the subject line] + scam."

5. **Consolidate**. Running a tight ship, with fewer accounts in general is a good strategy, and usually less expensive. Got domain names registered at two different registrars? Or three hosting accounts? Consolidate your domains with one registrar. Start with the one you like best, or the one with the most names. It's a bit of a slog, but there are always help guides and it boils down to having two browser windows open and being logged into each of the two accounts. Initiate the transfer starting with any domains that are not actively being used for websites to get the hang of it. For hosting, I like to get that mid-sized account, and use a domain not being used for an active site as kind of a file manger for the whole operation. I use HollyWorks.net as my filing cabinet, with several domains like HollyWorks.com hosted in that business account.

6. **Minify**. People may try to reach you via Zoom, Google Meet, Microsoft Teams, Slack, Skype, Facetime, Email, Text and Phone. Life is simpler if you resist maintaining every platform for communication, payment systems, and social media. Be selective. Pick two or three. There isn't an industry giant that hasn't been hacked, so maintaining a smaller footprint makes monitoring data breaches less taxing.

Web tools to keep handy

I keep a folder of links in my web browser for places I frequently visit. Over time it grows, and is occasionally weeded. The following are just a few tools that as a website manager you may need.

POCKET WEBMASTER DIGITAL ASSETS MANAGER

- https://docs.google.com/spreadsheets/d/1TqRZti2YQ2QKIRfZIlUUX6YwatSYFX5gkwnLX9efFcU/edit#gid=0

ACCESSIBILITY TOOLS

- Tota11y bookmarklet https://khan.github.io/tota11y/
- Functional Accessibility Evaluator https://fae.disability.illinois.edu/anonymous
- Accessibilitychecker.org https://www.accessibilitychecker.org/

AM I ONLINE? IS MY WEBSITE?

- Am I Online http://amionline.net/
- DOJ https://doj.me/
- Is is down right now? https://www.isitdownrightnow.com/
- Down for Everyone or Just Me? https://downforeveryoneorjustme.com/
- iplocation.net https://www.iplocation.net/

BROWSER EXTENSIONS AND GOOGLE CHROME

- Web developer toolbar https://chrispederick.com/work/web-developer/
- Google Lighthouse: in Developer Tools (view -> developer -> developer tools)

DOMAINS

- WHOIS ICANN https://lookup.icann.org/en/lookup
- Godaddy: Your Domains https://dcc.godaddy.com/domains
- Godaddy: Transfer Domain https://www.godaddy.com/help/transfer-my-domain-to-godaddy-1592
- DNS Checker https://dnschecker.org/

ECOMMERCE

- Sellfy https://sellfy.com/
- Podia https://www.podia.com/
- Teachable https://teachable.com/
- Payhip https://payhip.com/
- SendOwl https://www.sendowl.com/
- Gumroad https://gumroad.com/
- Merch by Amazon https://merch.amazon.com/landing
- Red Bubble https://www.redbubble.com/
- Teespring https://teespring.com/
- Shopify https://www.shopify.com/

Etsy	https://www.etsy.com/
Amazon	https://www.amazon.com/
Foodtruckr.com	https://foodtruckr.com/
Blogging Wizard	https://bloggingwizard.com/ecommerce-platforms-sell-digital-products/

EMAIL LISTS

Mailer Lite	https://www.mailerlite.com/

GOOGLE ANALYTICS ALTERNATIVES

PostHog	https://posthog.com/
Statcounter	https://statcounter.com/
Clicky	https://clicky.com/

WEB BUILDERS

Weebly	https://www.weebly.com/
Wix	https://www.wix.com/
Squarespace	https://www.squarespace.com/
WordPress	https://wordpress.com/
WPEngine	https://wpengine.com/
Kinsta	https://kinsta.com/secure-wordpress-hosting/

SMALLIFY IMAGES

Tinypng	https://tinypng.com/

SOCIAL MEDIA

Sprout Social	https://sproutsocial.com/insights/social-media-image-sizes-guide/
Facebook debugger	https://developers.facebook.com/tools/debug/
Knowem.com	https://knowem.com/

STOCK PHOTOGRAPHY

Fresh Folk	https://fresh-folk.com/
Women in Tech	https://www.flickr.com/photos/wocintechchat/with/25926827581/
Disability Resources	https://disabilityin.org/resource/disability-stock-photography/
Gender Photos	https://genderphotos.vice.com/
The Stocks.im	http://thestocks.im/

WEB SPEED

Pingdom	https://tools.pingdom.com/
Web.Dev	https://web.dev/measure/

WORDPRESS

Vulnerable WordPress Plugins	https://blogvault.net/vulnerable-wordpress-plugins/
WPWhite Security	https://www.wpwhitesecurity.com/blog/

Made in the USA
Coppell, TX
05 November 2026

11315631R00026